Windc

The New 2017 Updated User Guide to Learn Microsoft Windows 10 (user manual, tips and tricks, general)

PAUL LAURENCE

ISBN: 1544617070
ISBN-13: 978-1544617077

CONTENTS

Introduction

The new Microsoft Windows 10 features the back-to-basics start menu, your own virtual assistant, Cortana, a brand new built-in browser, an integrated version of Office for active users, and numerous other new functions and features in an attempt to make our lives easier to manage. It combines tablet and touch functions originally created for Windows 8, which run on a greatly improved operating system.

Windows 10 is also compatible with all Windows phones, Xbox, and most tablets. Microsoft's developers have anticipated Windows 10 to be the operating system for ultimately billions of devices. The Windows Store is full of a variety of apps ready to be downloaded for a multitude of personal uses and functional business uses that will sync with all of your devices.

The perks are wonderful, but with all of the changes, whether you are a user of Windows 7, Windows 8, or even Windows XP, you'll find yourself looking around to find the function or application you desire. That's when Cortana, the Windows 10 virtual assistant, can help you.

Chapter 1 – Activation with a Microsoft Account

You will need a valid email address. Microsoft prefers that you use their email service for your email address simply because they want you to use their products. However, activation can be done with any email address. Just be sure to write down your Microsoft account information someplace safe, as there will be many instances in which it will be needed for you to use and enjoy all the unbelievable features offered with Windows 10.

Microsoft Windows 10 activation will require some patience, whether you are just activating a new device, or upgrading from other compatible operating systems such as Windows 7, Windows 8, or Windows 8.1. The end result is a more secure method to verify that the operating system you are activating is, in fact, genuine. Going through this process links your Microsoft account to your user's digital license.

Sing a fingerprint instead of a password has been programmed into the Windows 10 "Hello" biometric platform. Basically, your fingerprint will be your password. Although, this new feature is built-in to the Windows 10 platform, it definitely depends on the make and model of computer you're using to run Windows 10. The new Windows 10 platform also supports not only fingerprint sensing, but also face recognition and even iris scanning. Newer computers in the marketplace today are building kits into their systems to support this important, highly secure log-on feature.

Don't be fooled! There are so-called free Windows 10 operating systems available for download on the internet.

CAUTION: These free versions are completely full of malware, trialware, ads, and possibly dangerous viruses that are not only annoying, but quite frankly, very unsafe.

Microsoft has fixed the bugs in early releases of Windows 10 that gave an error unless a user upgraded sequentially from Windows 7, then to Windows 8, then to Windows 8.1, and then to Windows 10. With these fixes, Microsoft ensures easier upgrades and clean installations, with the user being able to use previous version product keys. That's good news for upgrade costs.

A note about hardware changes to your device (i.e., replacing processor or motherboard) Windows 10 will likely need to be re-activated.

Entering the product key is available at several points during the setup and install of Windows 10, and users will be prompted to enter it, or you can skip those prompts to enter the product key later.

To input the product key from the desktop, follow the following procedures:

1. Click on the Start Menu

2. Click on Settings

3. Click on Update and Security

4. Click on Activation

5. Select Change Product Key

6. Enter the product key from your old copy of Windows (7, 8 or 8.1) or your new product key if you are doing a Windows 10 clean install

To manually link an account to a digital license:

1. Click on the Start Menu

2. Click on Settings

3. Click on Update and Security

4. Click on Activation

5. Click on Add a Microsoft Account

6. Sign into the Microsoft account you are linking to

Many users have had questions, especially regarding the new-release build product key usage. In my research, I discovered information on Microsoft's official website to be very helpful in determining the exact type of install that most users will be dealing with.

https://support.microsoft.com/en-us/help/12440/windows-10-activation

Chapter 2 – The Basics are Back

Microsoft touts Windows 10 to be “the best Windows yet,” and I completely agree. Since Windows 8 and Windows 8.1 weren’t well received by their user base (there had been a great deal of confusion with touchscreen devices), Windows 10 addresses the connectivity and recognition issues, but backed it up by going back to basics while providing superb features to optimize productivity.

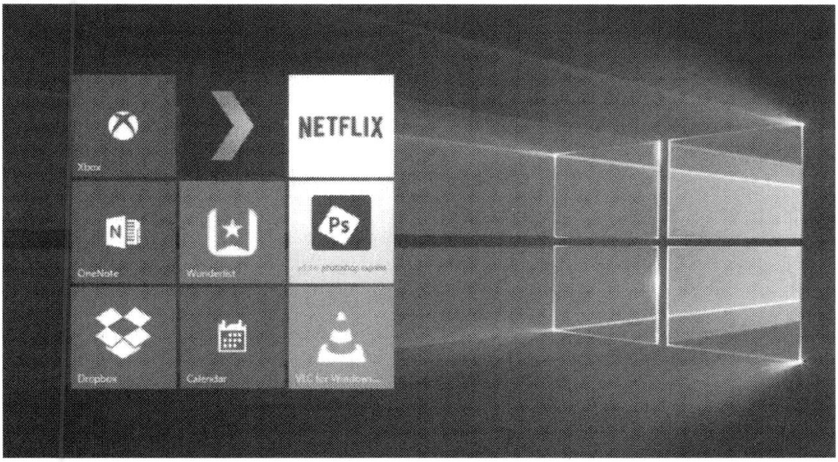

With the return of the start menu, users can navigate the interface and all installed applications with just a few clicks and no confusion. Windows 10 still possesses live tiles, which enables users to customize their desktop space with their favorite websites and apps, (i.e. email, calendar) all in one spot, just to the right of the start menu.

The Metro apps are back and in alpha default; they are available with one click of the start menu. You'll see all of your installed apps listed alphabetically on the left, and your live tiles on the right.

Win32 interface makes its return to launch apps, change their size, drag and drop, one-click close, and give you the assurance that your programs and apps are going to work the way they are supposed to. Users now have that familiar feel of the taskbar back. Pin apps to the taskbar with a simple drag and drop.

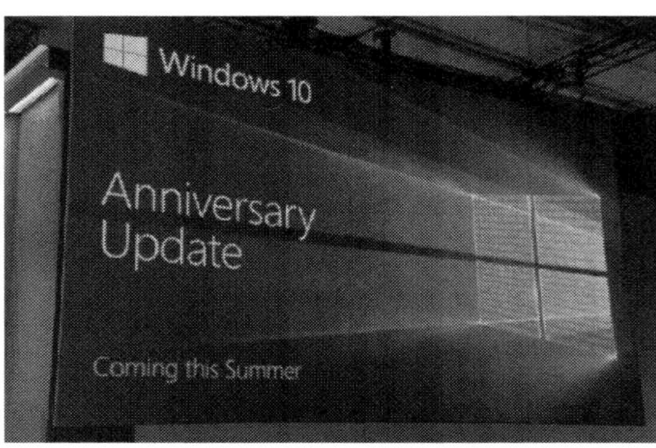

Speaking of pinning apps, Windows 10 has a great new feature that plenty of users are going to find very useful. You can now pin a Recycle Bin shortcut to the start menu and to File Explorer. Just search for the Recycle Bin from the taskbar, right-click on the link that appears, and choose **Pin to Start**. You can do the same from File Explorer, too. This new feature makes it very easy to keep your folders cleaned, especially when all you need to do is drag and drop from one folder to another while in File Explorer.

Also, right-click commands from the desktop give users numerous functional control.

Although File Explorer is part of the basic operating system, Windows 10 has put the File Explorer on steroids with substantial changes. The layout of the default settings might have you saying "Whoa!" Don't fear, you can change the layout with a few clicks and see the more familiar layout from Windows 7, Windows 8, and Windows 8.1. You must download an executable .zip file from Microsoft, unzip the file, and then upload it to your device. It only takes a few minutes and it is safe from the Microsoft website. Go to Microsoft.com, and just search for "changing File Explorer on Windows 10." Scroll down the page and you'll find information to change to the basic format along with instructions on how to make that basic format your default layout.

Most users find File Explorer quite easy to navigate once you really pay attention to the labels above the different file access areas. Research has found that most users report using the default mode because navigating using the right side is fantastic for copying and pasting, and dragging and dropping files and documents from cloud-based applications to the local drive.

Labels are clear and are just what they say. Frequent folders, for instance, is just that. They are folders that you have most frequently accessed. Recent files are documents that have been accessed in the past few minutes, hours, or days. The left side shows the actual name of the document, and the right side displays the path for the document within your local drive. These features provide instant access to documents you may be working on off-and-on during the course of a day. Now you don't have to search through the many directories.

Today's cloud services have fantastic apps, many downloadable from the Windows Store. Microsoft's cloud-based storage is available to Microsoft users and is called OneDrive. It's free for up to 15 GB of cloud storage space. It's simple to download OneDrive, and it will prompt you for your Microsoft account credentials. After download, it will appear under the list of directories to the left in the File Explorer. Some Windows 10 versions already have it in the directory for immediate use once you provide Microsoft account credentials.

Chapter 3 – Windows Platform & Universal Apps

Windows 10 users and developers especially love the new Windows Core, which is a single-unified platform that allows one app to run all of your Windows devices, such as your tablet, your laptop, your desktop computer, your Xbox, and even your cell phones. Windows 10 delivers this with their new Windows Core from just a compatibility aspect alone. Microsoft developers really heard and analyzed their customer base's relationship with their mobile devices. Mobile device use and sales have grown astronomically in the past 10 years and has led to a whole new level of app exposure and experiences. In today's market, users are using downloadable apps to pay their insurance, track their blood pressure, blog, and interact with social media. It's a whole new world of apps out there. Windows 10 has transformed into Windows Core and a one-developer platform, giving the user the same experience on all of their devices that run the Windows 10 platform.

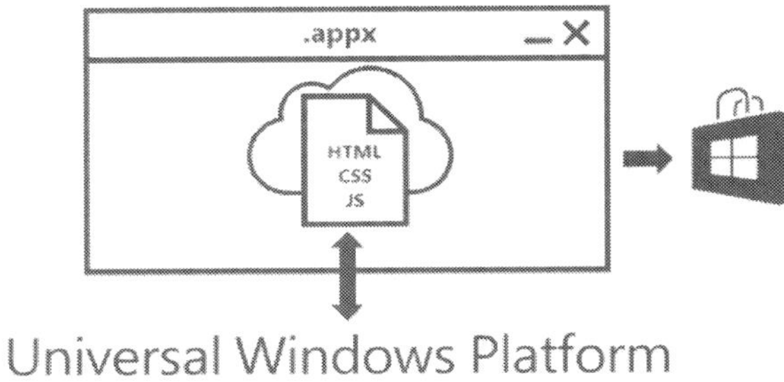

One of the biggest challenges that most apps have faced is supporting all the number of screen sizes available on mobile devices in the market today. Then, on top of that, they have to contend with touch, mouse, keyboard, game controller, and pen interface hardware, which is often difficult to go from app to app while maintaining a consistent page transition.

Quick transition is imperative for the customer that flows between devices for different types of computing. Windows 10 makes that very easy with the capability to run the same platform operating system on your laptop, tablet, desktop, and phone.

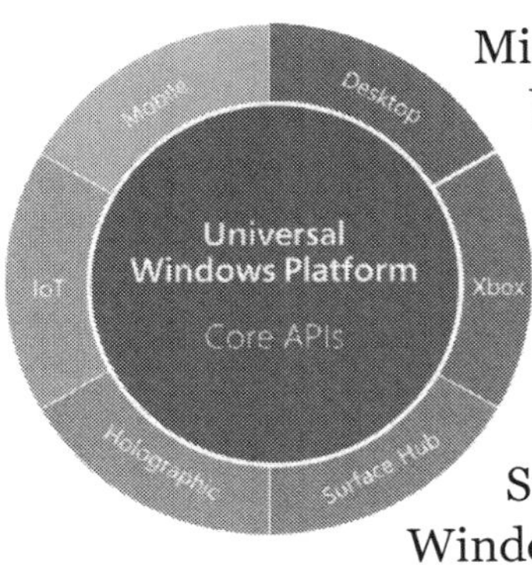

Microsoft, it seems, is trying to bridge the gap on the many devices its customers really want to carry. Research has shown an emerging trend to multi-modal devices, such as the Microsoft Surface Pro. The Surface Pro comes loaded with Windows 10, and is a hybrid between a laptop and a tablet, but also has modem and cellular phone capability. It's one device that can do it all.

Chapter 4 – Settings, the New Control Panel

The old Windows start menu and the Control Panel were available as a menu list or a drop-down list. Windows 10 allows you to pin the Control Panel to both the start menu and to the taskbar once you know where to find it.

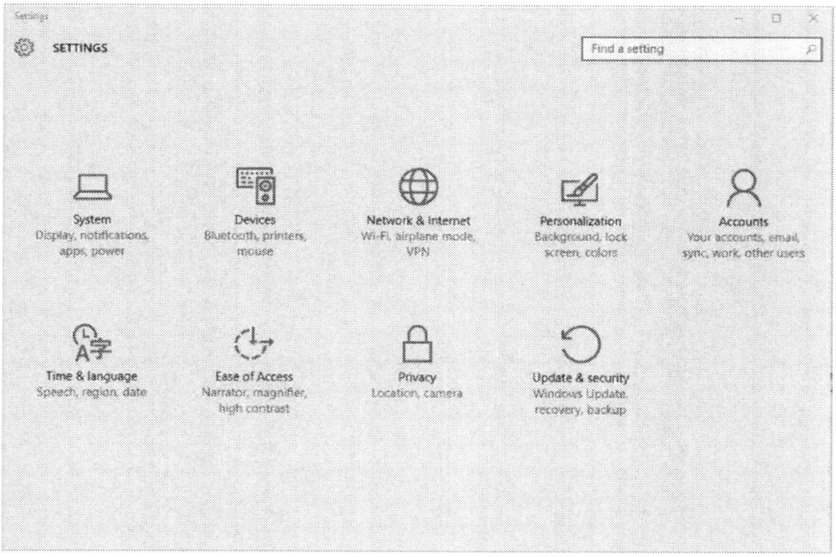

Windows researchers discovered that a major complaint

about Windows 8 was “where did so-and-so go?” due in large part because Windows 8 removed the start menu. Windows 10 brings back the capability of going old school with a start menu and Control Panel, making finding so-and-so a whole lot easier for users.

In Windows 10, finding the Control Panel is easy. Simply type “Control Panel” in the search box or click on the microphone to activate Cortana, and tell Cortana, “Control Panel.” More on Cortana in the chapter 6.

The Control Panel will show up on the right menu, and you can either click on it to open it from the search results, or you can hover over the icon, right-click, and choose “Pin to Start,” or “Pin to Taskbar” for ease of access in the future.

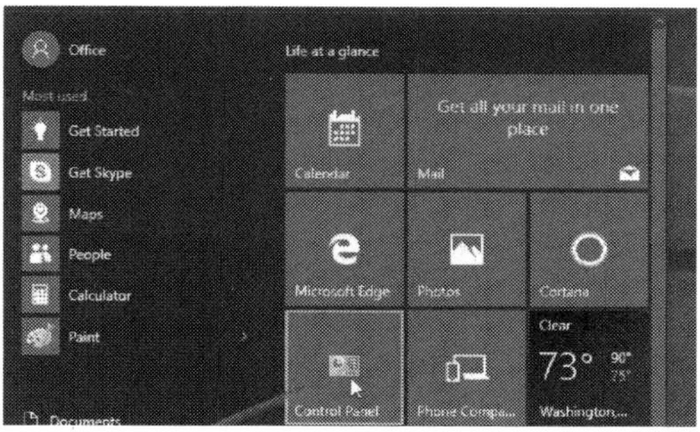

If for some odd reason, your search results for the Control Panel come up null, simply open the Command Prompt, and type in “Control Panel.”

That being said, the old school Control Panel has been integrated with Windows 10 settings. So basically, Control Panel is now Settings in Windows 10. Windows 10 settings are easy to access by simply clicking on the start menu. You will see a menu to the right. Looking down the menu toward the bottom is a sub-menu that indicates (from top to bottom) your Microsoft account login information. Under that is File Explorer (quick access). Under that is Settings and under that is Power. Click on Settings and you will see the Windows 10 upgraded version of Control Panel.

The use of Settings or the Control Panel is an important applet that helps the user make adjustments to pre-installed hardware and software that is quickly accessible on your Windows 10 device. The truth is that Microsoft Windows is eventually going to do away with the Control Panel. Using the Settings applets is far more comprehensive and easy to use and Microsoft wants to bring on a more touch-friendly Windows 10 settings menu.

From the Windows 10 Settings app, you have the ability to change most of your computer's basic settings. Windows 10 settings will look familiar to Windows 8.1 users, but it's now more dynamic. You will have immediate access to view, and/or modify the System, Devices, Network, Internet, Personalization, Accounts, Time and Language, Ease of Access, Privacy, and Update and Security. There's even a search feature at the top of the main window where you can search for specific applets, and anything that you type into this search box will prompt a drop-down menu of suggestions. Let's go through and discuss some of the more relevant features in the Settings/Control Panel.

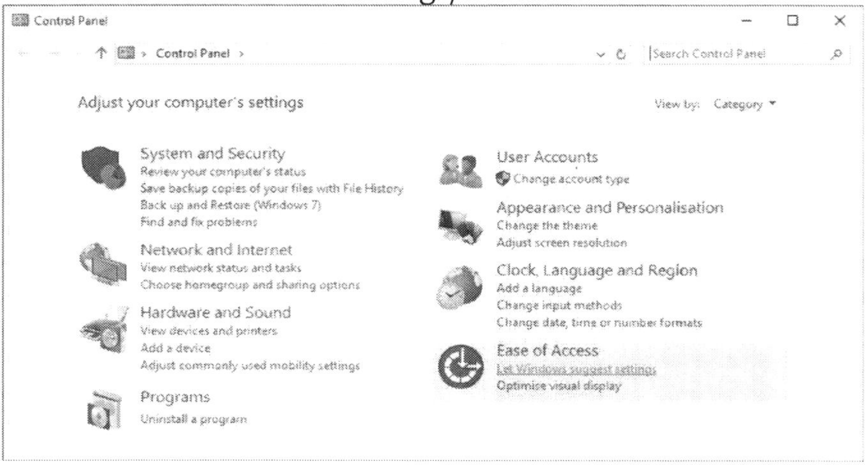

System: This contains your computer's general settings and setup utility functions. When in the System applet, click on About to see your computer's basic specifications, which also includes information about your computer's processor, RAM (memory), and operating system information, as well as information regarding the specific edition of Windows that you are running. From this System About applet, you can also change your computer's name by clicking Rename PC. Name your device to be able to easily identify your particular device on your network(s). It makes it so much easier to identify specific devices when sharing documents, files, and printers.

Display: The Display tab gives the user the option of changing your computer's basic options, including display settings, changing power options, and choosing default apps for the opening of different file types and protocols. To change your device's screen resolution, click **Advanced to display settings** to set up multiple displays and to calibrate your screen.

Tablet Mode: Windows 10 allows the switching between tablet-mode and PC-mode in a snap by enabling the tablet-mode on and off with the flick of a switch. Choose the way your computer handles sign-ins or go straight to tablet-mode, directly to PC-mode, or sign in to resume where you were when you last shut down your system.

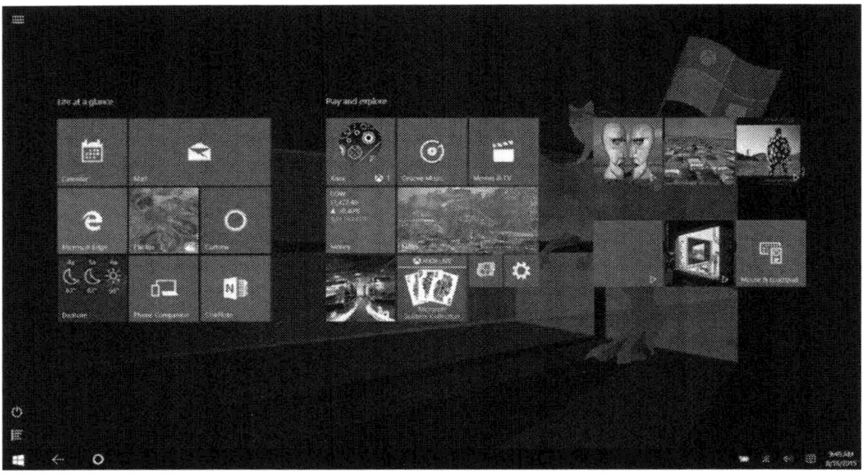

Storage: Checking your hard drive space is easy by simply clicking on Storage from Settings. The Storage screen indicates all of your device's storage drives, including any external storage or media drives you may have attached to your device at the time. In addition, you have the option to set the default save locations for documents, music, videos, pictures, apps, webpages, etc. For example, you have received secure document(s) from a reliable email source. You would just click on the download feature within the email content, and based upon the defaults you have chosen, the document(s) will download into that specified directory. Setting download defaults is a great way to consistently remember where your downloads are located on your device. To choose and save a download location for your file types, pick a location from the drop-down menu, then click apply.

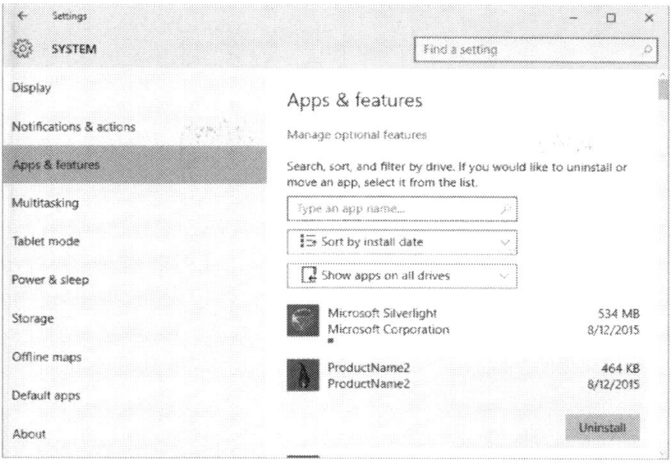

Apps and Features: Click on Apps and Features from Settings to see an alphabetical list of all installed apps on your device. This menu allows you to sort not only by an alphabetical list, but also allows sorting by date, sorting by file size, and sorting by installation date. Apps and Features allow the modification and uninstallation of any app shown on the list.

The Settings menu is an important tool in managing the apps and specific settings for your device, and to tweak advanced settings. It is recommended that any new Windows 10 user click around in the Settings menu to familiarize themselves with all the options available there are in addition to the important features listed above. Microsoft includes links throughout the Settings menu that will redirect to the Control Panel, if necessary, so locating a specific setting isn't difficult at all.

Chapter 5: Task View for Virtual Desktops

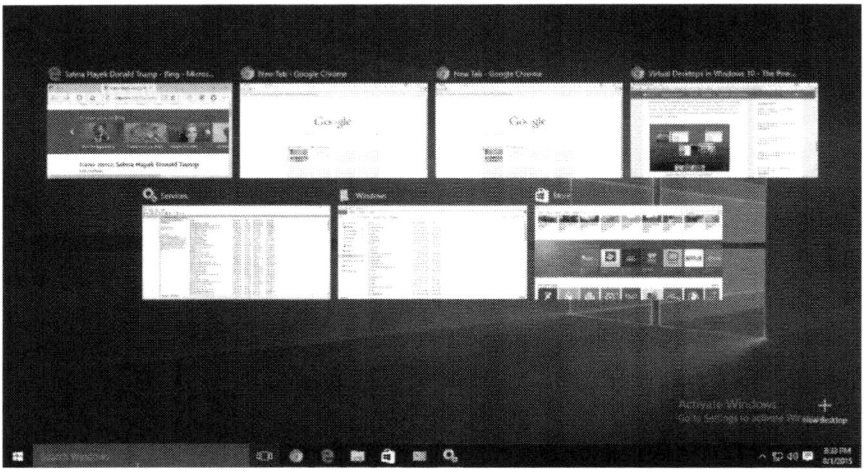

Windows 10 *finally* launched virtual desktops as part of their mainstream operating system. A virtual desktop is an integrated applet that runs from Task View. Make no mistake about it, it can create a powerhouse of productivity for the power user. Keep in mind that Task View has a different approach than previous versions' way of navigating through open windows with the keyboard shortcut Alt + Tab. Task View shows all open windows that stay in permanent view until you close, or click, a window to be in the background.

Click on the Task View icon, next to the Cortana/Search entry box located at the bottom left on your taskbar to enter Task View. When you click on the Task View icon, it displays your open windows, a virtual desktop, which makes it really, really, easy to return to programs or documents for those times when you have ten windows open at once. This is particularly useful in the business environment when it is sometimes necessary to use a multitude of documents to create a new document. You can cut and paste from several different documents with ease in the virtual desktop without any guess work.

Also, for multi-monitor users, Task View will display on all your monitors the same when you click on the icon, then you can click through the display pages of Task View on each respective monitor to display different display pages of all your windows open in Task View.

Let's talk about Task View using Snap. Since Windows 10 supports Snap, you can set a window to take up half the screen. And, with the latest update to Windows 10, you can now use Windows 10's newest feature named Quadrants. When the function is called up, Quadrants sets your programs into a four-rectangle grid on your screen.

Using Snap while in Task View is easy, just click the Windows logo key and use one of the side arrow keys. Two of the side keys snap the window to the respective half of the screen. Quadrant executes and snaps the window to the bottom or top of that side's half.

This is beneficial when you have multiple applications open, and/or even multiple desktops.

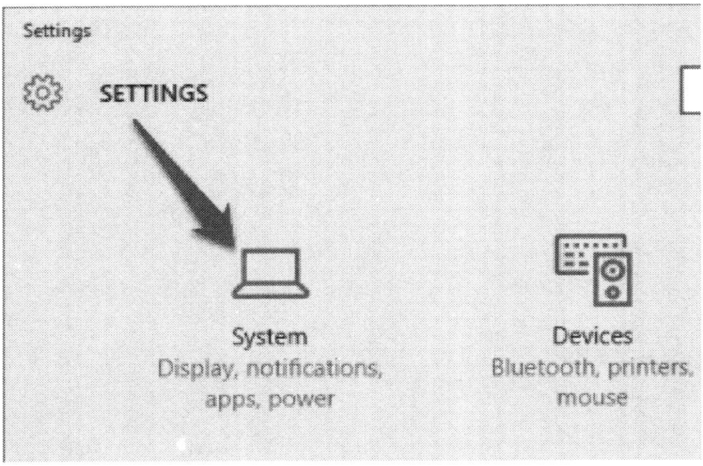

Windows 10's newest release has an emphasis on keeping its users organized, and virtual desktops are just one great example of organization. Think in terms of organizing your virtual desktops in categories of substance. Say you are a web developer with graphic design services with a flair for music. For example, you could set up a virtual desktop for all html, xhtml, asp, php, and javascript documents under "Desktop 1" virtual desktop, then set up all your graphic design programs, documents, and files as "Desktop 2" virtual desktop, and all your mp3s, YouTube links, and games for those quick breaks set up as "Desktop 3." It appears that Windows 10 has no limit to the creation of virtual desktops. This user has had up to 15, and my system still performed at its optimal level.

Creating a new virtual desktop is easy! Just click the Task View icon located in the taskbar. Once the Task View interface loads, click on the + New Desktop link in the lower right-hand side to create the new virtual desktop. Regardless of the desktop, you are located on, Windows 10 defaults to showing only open, active windows and programs. To change these settings, go to Settings > System > Multitasking > Virtual Desktops to change these settings.

New virtual desktops are, by default, created to the right of the line. You can use your right arrow key to navigate to the last desktop, and the left arrow key to navigate your way back.

Additionally, you might consider using a few keyboard shortcuts. Simply hit the Windows logo key + tab. Then hit the Windows logo key + D to create a new virtual desktop; and Windows logo key + Ctrl + F4 will close the virtual desktop you are on.

So this writer has a personal tidbit to share with you. I often access my virtual desktops using keyboard shortcuts. I will warn you: it's very easy to hit the Windows logo key + Alt + arrow keys. I have done it many times. Really! The result of this minor faux pas is the instant changing of your screen display orientation. If this happens to you, no fear, you have been warned. To fix it: simply hit the Windows logo key + Alt + the up arrow key to resume your normal landscape mode.

Chapter 6 – Cortana Has Ears

Windows 10's Cortana is perhaps the most notable and visible new feature. Cortana is like your virtual personal assistant, and the more you use Cortana, the better, and more personable the experience.

To access Cortana, locate the search box. It should be in the taskbar area and have the words "Ask Me Anything" on it. Simply type a question in the search box and the results will pull up in Cortana's menu. You can click on whatever topic or search result you were inquiring about. In addition,

Cortana is voice-command capable, so yes, you can talk to Cortana if your device has a microphone. To activate Cortana's voice command, simply click on the microphone icon located inside the "Ask Me Anything" search box.

Cortana is pre-programmed for many commands, and here's a list of some of the most used commands, but as I mentioned earlier, the more you use Cortana, the more Cortana gets to know you and rewards you with a more personalized user experience.

1. Message and play games
2. Send emails and texts
3. Creation and management of lists
4. Manage your calendar and keep you up to date
5. Give you reminders based on time, places, or people
6. Track packages, teams, interests, and flights
7. Open any app on your system
8. Find facts, files, places, and info

Also, if you have a Windows phone, you can sync notifications between your PC and your phone.

To set Cortana to automatically recognize your voice command, click in the "Ask Me Anything" search box to open Cortana, then to the right, click on the Settings menu for Cortana. Where you see Let Cortana Respond to "Hey Cortana," toggle the setting to "On." It makes for a completely hands-free experience. And quite frankly, it's so convenient and fun.

For the optimal experience with Cortana, users should devote some time setting up Cortana's Notebook. To access Cortana's Notebook, simply click in the search box on the taskbar to open Cortana, then click on Notebook. Here you will find a menu list to tell Cortana about your interests, favorite places, news, events, weather, travel, sports, and customize other events you want to keep up on. The more you add, the better Cortana can serve you.

Cortana is also available to serve you while you are browsing the internet. Cortana and the Windows 10 built-in browser, Microsoft Edge, will take you to new heights with your online experience.

When you are visiting websites that Cortana supports, Cortana will show up in the address bar and list suggestions to further your research or further your search for a specific person, place, or thing. Just click on Cortana's message for Cortana's assistance. For example, let's say you are on PapaJohns.com contemplating an order from their website. Cortana will point out directions to stores, contact information, menus, and hours of the store.

Also, another example of a great use of Cortana and Edge is saving time and money while doing online shopping. For example, let's say you are on Macys.com. Cortana will make suggestions of items on sale and alert you to coupons that are available to use at checkout, and/or on in-store purchases. Just look for Cortana's circle in the browser taskbar, click on it for coupon and discount suggestions, and/or coupon codes for discounts on your online purchase.

In addition, use Cortana to pursue further research on any topic that you want or need to know more information about. For example, you've done a search for elephants. You will see website after website on your search results offering information regarding elephants. Hypothetically speaking, say there is reference to a species, or just a word that you don't understand. Just highlight the word, hover over the highlight, right-click, and "Ask Cortana" for more specific information regarding the term highlighted.

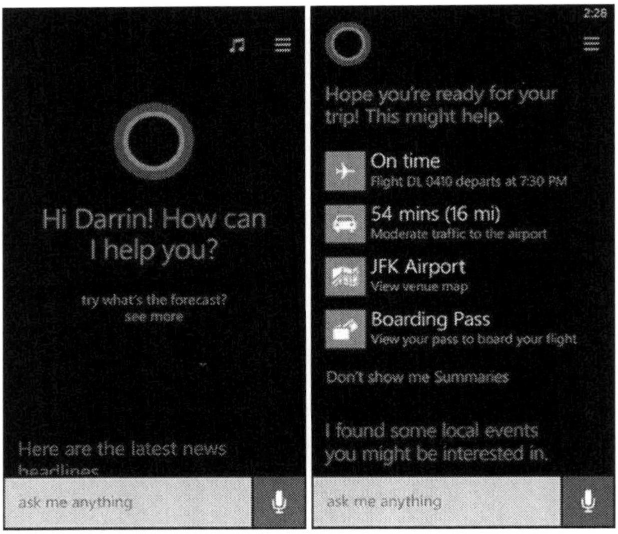

Currently, Cortana is powered and run by Bing and when you ask Cortana something that can't be answered, Cortana will automatically open your default web browser and a Bing search is performed. Cortana will open in Chrome or Firefox as well, but will always perform and display the search results using Bing.

Cortana is your virtual assistant and you can ask for help at any time. Just say, "Hey Cortana, help." Cortana will display a comprehensive list of options available to you. Microsoft is adding new features and improving existing ones in free updates moving forward.

Chapter 7 – One Drive Cloud-Based App

Windows 10's OneDrive is integrated into the operating system and is an important segment of Microsoft's all around platform. Valid Microsoft Windows 10 accounts are automatically set up for OneDrive when you log in with a Microsoft account. You will find free cloud storage easily accessible in the File Explorer, and when synced, OneDrive maintains your documents and files on a cloud storage server. Just look for the blue cloud icon for OneDrive.

There are some new features and upgrades from Windows 8.1 regarding OneDrive. Users are offered more flexibility and cloud storage has become much easier to navigate, sync, and understand.

When it comes to cloud storage, much like any content delivery service, there is something for everyone. Microsoft's OneDrive is a Windows 10 user's best choice. Like all services, OneDrive has its good points and bad points. Believe me, those good points are good and deserve your attention.

Believe it or not, it's not just limited to Windows 10. OneDrive can be downloaded on your Android device, iPhone, or iPad. No matter what you use, you can experience the convenience of a free cloud storage service!

OneDrive and Office 365 work in total harmony and 15 GB of free space is great, but if you also subscribe to Office 365 you'll get even more. Part of the Office 365 package is a 1 TB OneDrive storage allowance for as long as you subscribe. Even the power users of Office will struggle to fill that. It's a great place to store your music, pictures, and videos to save space on your local drive. OneDrive is also integrated with Microsoft Office so when you're using the desktop apps, mobile apps, or the web apps, your documents will always be in sync.

OneDrive streams your own music through the Groove Music App/Xbox. Now, Windows 10's latest build allows you to upload all your recorded streams, mp3s, mp4s, and videos that you can then stream through the Groove Music app. The Groove Music app does have some instability bugs, but nothing's perfect. It does, however, give you cloud-based access to your personal media collection.

To change OneDrive settings, right-click the icon in the notification area, select *Settings,* toggle to the *Choose Folders* tab, and click on the *Choose Folders* button. You can then sync *all files and folders on your OneDrive,* or *choose individual folders to sync*. This means they will be available on your local drive as well.

One Drive is integrated with Cortana and other universal applications. Cortana can answer questions about OneDrive, can search your OneDrive files, including the ones stored on the OneDrive cloud that may not be synced at the time of the search. Of course, that is provided that you are properly signed into your Microsoft account. Cortana will bring any matching files to your request in a menu for you to select from the search results. Cortana can be activated from voice control and taskbar inputs.

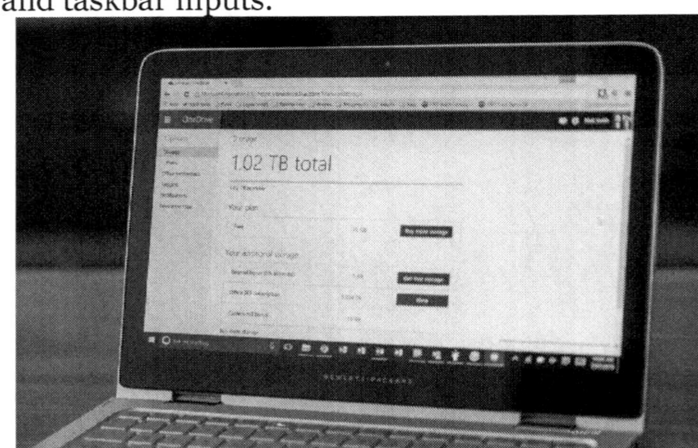

OneDrive can aggregate large photo collections that may or may not be spread across multiple devices. The new Photos app utilizes OneDrive to pull images from all your devices. It can remove duplicate images or similar images from its collection, can automatically enhance photos, and create photo albums based on time, people in pictures, and place.

In the latest updates for Windows 10, and specifically the OneDrive platform, Microsoft has introduced the new Universal Windows Platform (UWP) app for OneDrive that works on PCs. You can download it for free from your device's Windows Store.

After you have downloaded and installed OneDrive from the Windows Store, sign-in to the app with your Microsoft account, and your OneDrive files will populate. When all your files have populated, you can offload some of your large files to OneDrive to save storage space on your device. You will still be able to see them on your device; they are just not stored on your device. In fact, once you get familiar with using the app, you will likely choose to leave all of your larger files on OneDrive which keeps your PC or another device computing at its optimal level.

Offloading large documents, files, and folders to OneDrive is easy. So, if you wanted to leave your backup folder in the cloud, but you don't need or want it taking up space on your PC or another device, then you can click on the upward-facing arrow to the far right of the taskbar, then right-click the OneDrive icon in the panel that appears. From the context menu, choose Settings. Uncheck the box next to

Backups. You will see a warning window pop up letting you know that your files will be deleted from your PC but remain stored in OneDrive online. Then click OK on the popup warning window, and close the remaining OneDrive settings windows. Offloading your files to OneDrive is easy and your files are just a click away when you need them. Of course, you can do the same thing from the web app, but using the Windows Store app is more convenient.

OneDrive also supports remote access. From the Settings tab, if you check "Let me use OneDrive to fetch any of my files on this PC," you can access your files remotely, i.e. from another computer via the OneDrive website.

When you enable this option, the connected PC shows up on the OneDrive web interface. You can access the online interface by visiting onedrive.live.com. If the respective computer is turned on and connected to the internet, you can access any of its folders, files, images, videos, etc.

Sharing folders and files with others is easy in OneDrive. In the original release of Windows 10, sharing from the OneDrive local folder in Windows 10 was not possible; however, providing your system is up-to-date with the latest version of Windows 10, it is now possible to obtain sharing links to copy and paste into emails, etc. to enable collaboration with others.

In my opinion, the web interface is much more functional and you can login at onedrive.live.com. Browse the folders and/or files until you find what you want to share; then right-click, and click "Share from the Context Menu." You are then given an opportunity to assign viewing and/or edit privileges simply by entering the recipient's email address. Note that once you share a folder or file, the shared icon appears in the lower right-hand area. Also, just right-click the shared folder or file to open the Share Menu to revoke any privileges.

For many years now, cloud storage has been a way to share files with others or transfer them from one device to another. As cloud storage space becomes increasingly more affordable, cloud storage has become an increasingly attractive means to store backup data if you don't mind someone else managing your data. This latest update for OneDrive now syncs across many platforms and amazingly integrates with apps you most likely use, such as Outlook and Office.

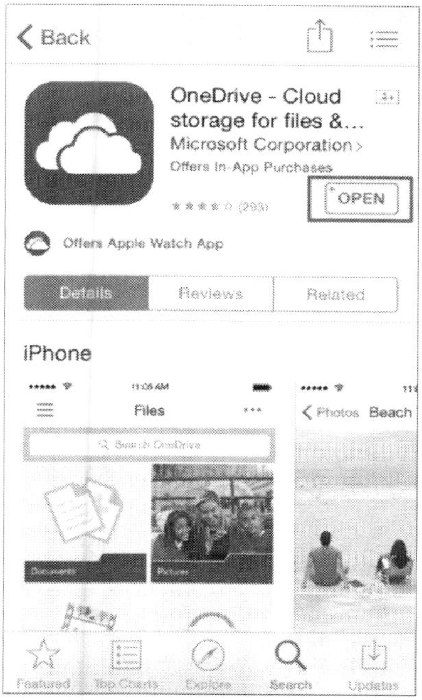

Like most applications, OneDrive operates on a by default setting that can be adjusted or turned off. OneDrive will work on iOS and Android, but you must install the OneDrive app from your device's store to sync your OneDrive. In my research, it has come to my attention that Microsoft is planning updates with more magical features to make sharing from all your devices with friends or business associates much easier, and will be published and available sometime mid- to late 2017.

Microsoft is heavily promoting OneDrive to sync and store the data from across all of your devices. And they are offering free cloud storage for up to 15 GB, but with an Office 365 subscription, you get 1 TB of online storage that seamlessly integrates with Windows 10 and many other Microsoft applications. It also is reassuring that all your data is automatically synced, backed up, and most importantly, easy to access. Just remember to be safe with your backup data. This writer follows the 3, 2, 1 backup method. That's 3 copies, on 2 types of media, with 1 copy offsite. With that data mantra, you can preserve your digital records making up your life for many years to come.

Chapter 8 – Mail App & Calendar Improvements

With the latest updates from Microsoft, the Mail and Calendar apps in Windows 10 now have lots of new features including the Focused Inbox that was previously introduced in the mobile version of Outlook. It seems similar to Google's tab option. Microsoft insists that the Focused tab emails are a dissemination of the people you commonly interact with, but doesn't offer any specific criteria for training the Focused tab.

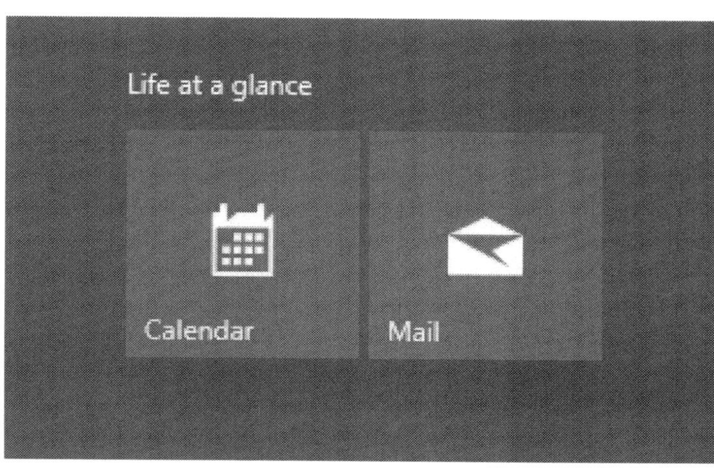

Lots of architectural updates speed up both the Mail and Calendar apps' performance and are very similar to that of Google's Gmail and Calendar tools. With this latest version, it's much easier to set up various mail accounts so you can receive all your mail in one place.

Updates to the Mail app also offer benefits from the addition of @mentions. You can very quickly add contacts in the To field, and it provides an easy way to filter messages based on referral.

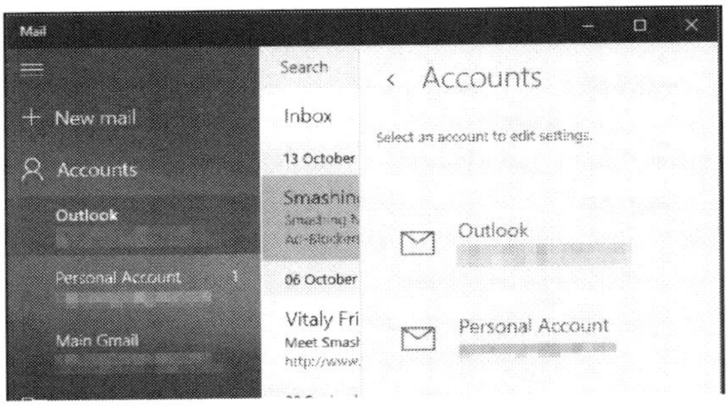

Windows 10 comes with the new Mail app pre-installed and it's a greatly improved version from the one in Windows 8.1.

Once you are logged into your PC or another device with your Microsoft account, the Mail and built-in Calendar apps will automatically link. If you are logged in with your local account, you first need to setup the email account in the Mail app. Simply click on the Windows icon on you PC, then click on the Mail app's live tile.

The Windows 10 Mail app is not just about an Outlook account; you can add all of your email accounts and all of your emails will flow like Outlook listing the respective accounts for ease of navigation. Just click on the Settings icon in the lower-left corner, then click on Add Account, and follow the steps to create any of your email accounts. The Mail app will/can set up your email account automatically detecting your webmail account settings. However, from time to time, depending upon your security settings in your webmail account, you will either need to change the webmail settings, or provide Imap credentials to integrate the webmail account into your Mail app, similar to Outlook.

Backing up your email has never been easier using the Mail app. You can save or backup email messages as follows. Open the email app, then click on the 3-dotted Actions menu in the top-right corner, then click on the "*Save as*" link. Your File Explorer will open and you can choose where to save your backed-up email files.

Windows 10 Mail app now allows you to personalize your email by adding a signature for each and every email account you have created. Just click on Settings, then Options. Scroll down to the Signature tab. Click it "On," then add the signature you want to be shown at the end of every email you send.

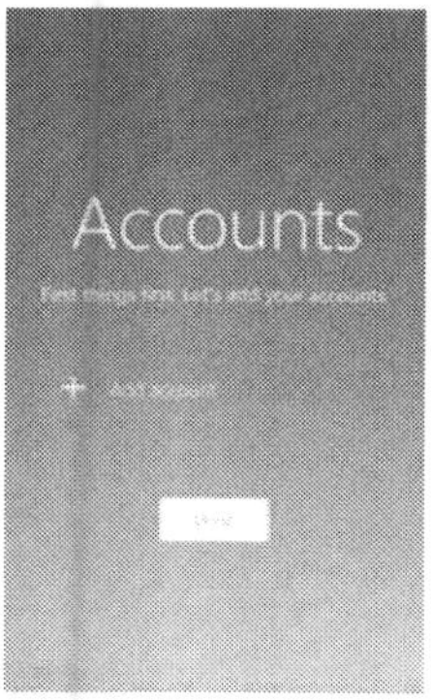

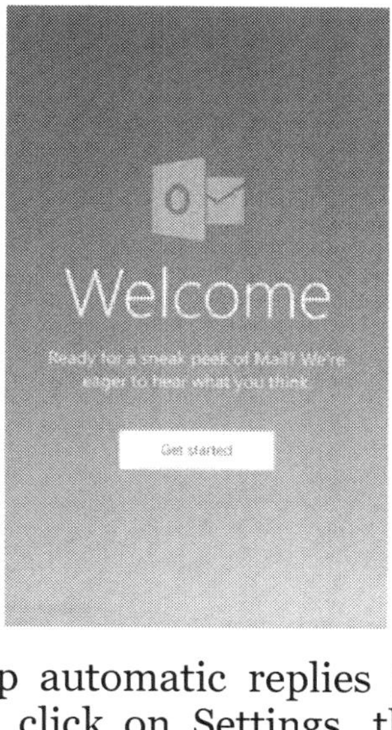

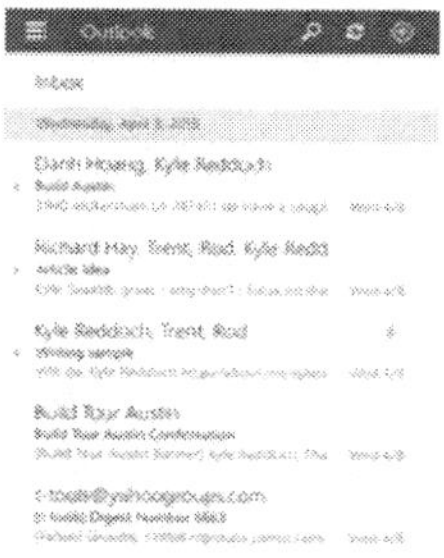

Also, you can setup automatic replies in the Mail app of Windows 10. Just click on Settings, then Options. Scroll down to the Automatic Replies tab where you can turn the option On or Off. You can also add a message you want to be sent with every automatic reply. You have the option to check the box just below if you want to send those automatic reply messages specifically to your contacts.

Now, let's talk about how the built-in Calendar app plays an important role in the productivity of the app. Type Mail in the search bar, then open the Mail app, then click on the Calendar icon in the lower-left corner of the app. Note the detailed interface with the birthdays of all your mail contacts, lists, events, holidays, and so much more. You can toggle the Calendar view to display in days, weeks, or months. This version of Calendar lets you create a new event and to keep the reminder as well.

When you open the Windows 10 Calendar app, all of your events from your Microsoft account are displayed. To display events from other calendars, add the accounts to the Calendar app. Just click the Settings button which will open a panel on the right. Click Accounts and all of your accounts will display. Click on **Add Account**, then a dialog box will display a list of the available services you can connect to the app.

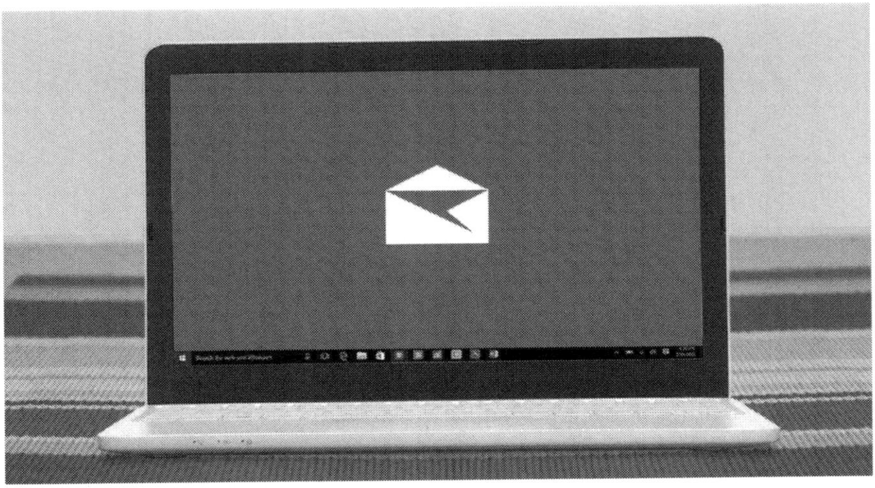

The look and feel of the Calendar app in Windows 10 have changed. The change that stands out the most in the Windows 10 Calendar app is the ability for users to sync and view their Google Calendar. Older Windows platforms did not support this feature of integrating your Google calendar.

To appeal to the eye, the Windows 10 Calendar app offers the ability to customize the background picture with just a few clicks. Click on Settings, click on Background Picture, then browse and select any picture available on your PC, OneDrive, Google Drive, etc.

Updates to the Calendar app have introduced color categories for an easy, visual interpretation of your schedule. Follow internet calendars, cable shows, and sporting events together with your appointments. Travel plans are highlighted which makes them easier to catch your eye, and Microsoft has made other improvements to calendar invites.

Windows 10 Calendar app will let you keep on top of your travel plans as well as package deliveries by adding simplified summary cards to your inbox and calendar. **Travel reservations and package deliveries** work across both Mail and Calendar. This allows you to quickly get to travel reservations and package delivery details, check in for flights, change hotels, change rental car reservations or track your package's delivery status with the touch of a button, which can all be set up with reliable reminders.

If you are a Skype user, you will enjoy the small improvements in the Windows 10 Calendar app to attend online meetings with an integrated Skype feature, and enjoy the use of the scheduling assistant tools integrated with Office 365.

Here's a link to update your Calendar and Mail from the Windows Store:
https://www.microsoft.com/en-us/store/p/mail-and-calendar/9wzdncrfhvqm.

Chapter 9 – Powerful Command Prompts

The Windows Command Prompt is an often overlooked, but powerful tool. It's reminiscent of the days of DOS and the command line, and has continued to be featured in all versions of Windows. With the passing of time, Command Prompt has become incognito while other exciting apps and features are introduced. I urge you to not ignore the Command Prompt, as it may be your best friend when troubleshooting your PC or another device. Windows 10 has introduced some handy upgrades to the Windows Command Prompt console window, making it a whole lot easier to use.

Command Prompt has some new tricks with the latest version of Windows 10. Resizing the console window is easier and you no longer have to mess around with console buffer size settings. The Command Prompt console now supports word wrap which means text doesn't appear off-screen, forcing you to scroll horizontally to read long lines of text. And, with a click, you can select console output with your mouse, and right-click paste to your clipboard.

Let's examine the armory behind the Command Prompt and advise you on how best to use it.

The most popular Command Prompt tool is the System File Checker (SFC), and by typing **sfc / scannow**, your system will do a scan to detect corrupt system files, and once detected, replace them with backups.

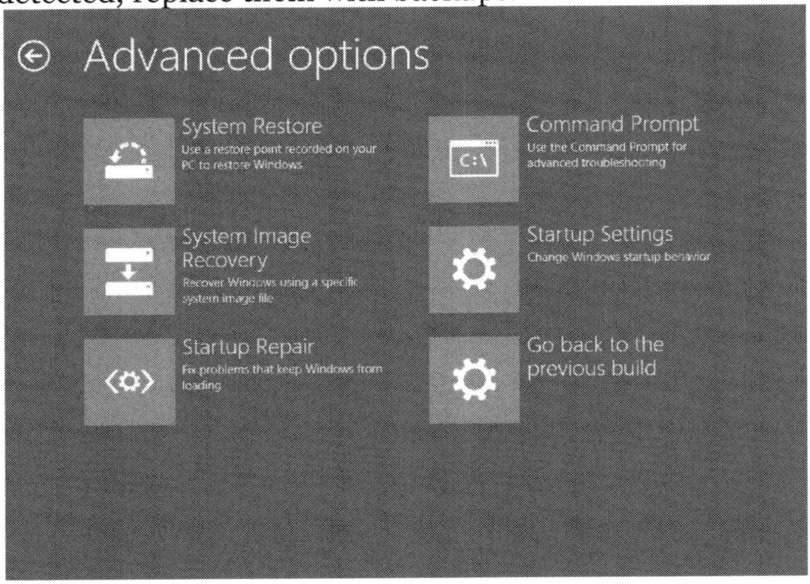

Windows 10 Command Prompt now allows customization of its appearance. Just right-click the Command Prompt icon on the menu bar, and select Properties. You'll see options displayed over four tabs: Font controls the text type, size, and style. Colors allow you to change background and text colors, plus, it makes the window semi-transparent.

An option, which is the main tab, allows you to tweak the command buffer, or command history, then you can move through previous commands using the up-cursor key. Also, you can easily change the mouse cursor size, enable or disable various new features which are new to Windows 10, such as copying and pasting text using keyboard shortcuts.

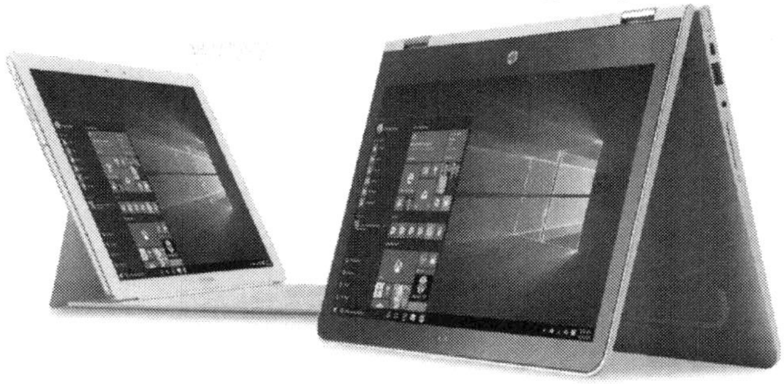

Layout tab enables you to set the window size in lines (height) and text characters (width), while the Screen Buffer Size height shows the number of lines you can scroll through. The default size is usually sufficient, but make sure you check the “Wrap text output on resize” option.

Windows 10 re-introduces some efficient features to the Command Prompt, including the enhanced Quick Edit mode (originally introduced in Windows 8).

This enables you to select and copy text from the console quickly, using the mouse. Click and drag to select, then right-click to copy it to the clipboard. Place your cursor to start where you wish to paste it in the console and right-click again.

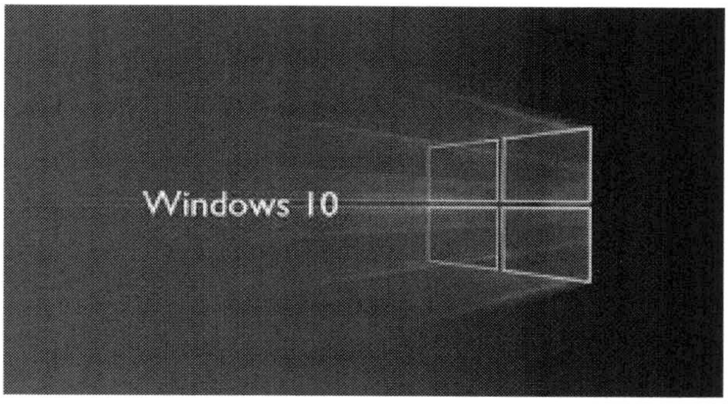

PowerShell features remain the same and allow you to switch back and forth between the two command-line environments and have all the features remain consistent. To see all the new PowerShell options, just open the Command Prompt and type “help” for all the PowerShell commands and their respective potential results.

Chapter 10 – Game Mode and DVR function with Video Capture

Windows 10's, brand-new Game Mode feature enhances the PC-gaming experience. The new platform minimizes your operating system's resources to almost nothing, allocating most of your system's resources to the Game Mode, which results in faster-running games and overall more stability, boosting Windows 10's gaming credibility with features that have proven to be very popular with the Xbox One. Windows 10's Game Mode almost solely focuses your GPU and CPU on the Game Mode–enabled process.

Xbox One users have said this feature works almost the same as Xbox One runs a game. Xbox OS allocates all or most of its resources so the game runs the best it can. Windows has never had a built-in Game Mode, so literally, the game is changing.

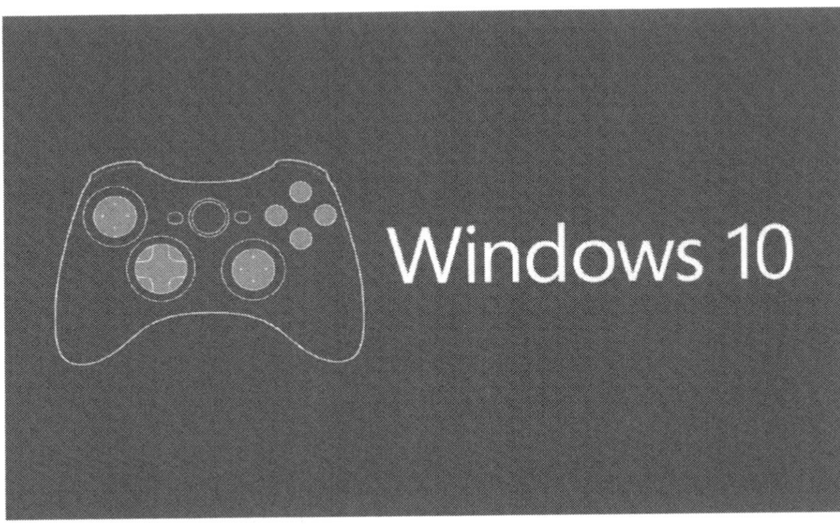

Microsoft touts that the goal behind Game Mode is consistency, rather than performance boosts (although there is that too). Game Mode basically prevents system tasks from taking resources from your games, allowing performance and frame rates to be generally more consistent. This will specifically enhance scenes and situations in your game play that are normally more intense on your system's hardware.

Microsoft also says that "while Win32 PC games will see some benefits from Game Mode, it will be UWP games that see the biggest improvements."

The reason for this is the UWP environment, which is a little more standardized than Win32. Windows 10 has optimized popular hardware configurations that include Intel, AMD, and NVIDIA, so games compatible with these hardware configurations are ready to use out of the box.

To enable Game Mode, simply flip the switch via the Xbox Game Bar found on Windows 10 by pressing the Windows logo key + G. Windows 10 will remember which games have been enabled until you turn it off. From the Game Bar, you will find a Settings feature. Click on the Settings feature to set up your gaming preferences, shortcuts, and audio preferences as well. You can also access your Xbox game menu.

Now, let's talk about the integrated DVR built-in to the latest version of Windows 10. There are many reasons why a built-in DVR is advantageous to any Windows user; however, from my research, the DVR is most widely used to allow gamers to record video from any game running on their computer using the Game DVR feature. Game DVR runs in the background while someone is playing a game and keeps a running video buffer so you can rewind and capture cool moments.

If something interesting appears in the game, users will be able to press the Windows logo + G keys for the recording menu, click on the video from the buffer, or start recording something in their game that they're about to do. The video files are saved as 1080p resolution MP4s, which can then be edited by most video editing software.

Video capture is widely used to record video evidence of your brightest gaming moments, but it'll actually let you create videos on any open app or desktop software (except for OS-level areas like File Explorer or the desktop). To summon it, simply press the Windows logo key + G. A prompt will ask you if you want to open the Game bar. Hit the "Yes, this is a game" option and different variants will show up in a floating bar. Just hit the circular record button to stop a video. You can find your saved videos in the Game DVR section of the Xbox app, or inside your user video folder which you can see under Videos, then click on Captures.

As you minimize or set your game to the background, Game Mode will disable itself, and give your system all the access to your hardware when you decide to multi-task.

Chapter 11 – Windows Updates Features Peer to Peer Downloads

Windows Update has seen a lot of changes to Windows 10. One of the biggest changes is a more aggressive approach to keeping everyone's Windows up-to-date. Windows 10 now uses a BitTorrent-style, peer-to-peer download for updates. Please keep in mind, though, that many of the included, universal apps like Microsoft Edge, will be automatically updated through the Windows Store, which is separate from Windows Update.

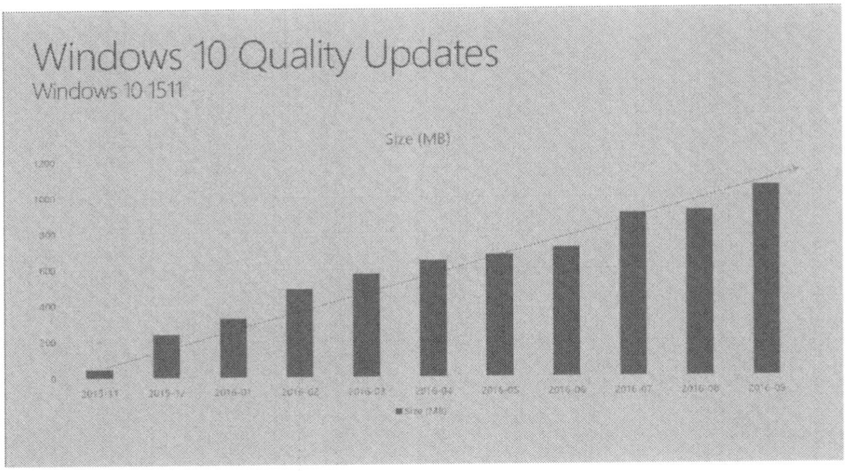

What are peer-to-peer downloads? Windows 10 has initiated peer-to-peer downloads for Windows updates. If you have several Windows PCs, you don't need to download the same update on all of those devices. You would merely update the first device, and the other PCs or devices could download it from the first PC. This scenario only works if you have your devices synced and your network settings set up to share information across your devices. To control peer-to-peer downloads, click on "Advanced Options" and click on the "Choose how updates are delivered" link. Also, by default, Windows 10 enables peer-to-peer downloads over the Internet. Your PC or your device will use your upload bandwidth to send those Windows updates to other PCs and devices. You can disable this by enabling the setting, "PCs on my local network."

A note about running Disk Cleanup which can eliminate your Windows update files on your PC. Assuming you are trying to free up space on your PC or device, just be aware that your PC won't be able to provide peer-to-peer downloads for your other PCs or devices because the files will have been removed due to Disk Cleanup.

Although Windows 8 had dual interfaces for Windows Update, the PC Settings app, and one in the older Control Panel, Windows 10 has retained most of the old Control Panel, but the actual Windows Update interface has been removed. The only Windows Update interface in Windows 10 is in the new Settings app under Update & Security. Updates install automatically, and you can no longer disable, delay, or choose which Windows updates to install. All updates, from security updates and Windows Defender definitions, updates, and driver updates, etc., are now installed automatically.

The only option you can control is the "Advanced Options" link and uncheck "Give me updates for other products when I update Windows." This allows you to disable updates for

Microsoft Office and other Microsoft programs installed on your device.

Visit the Windows Update interface and you'll only find one button: "Check for updates." Click this button and Windows will check for available updates. If it finds any, it will automatically download and install them. Windows will also check for updates in the background and automatically download and install them.

Windows 10 will not download updates on connections you have checked as "metered." This prevents Windows from wasting valuable tethered data or other mobile data on updates that can wait for an unrestricted Wi-Fi network. To prevent Windows from downloading updates on a specific connection, first connect to that Wi-Fi network.

Windows 10 Home users cannot delay upgrades at all, but Professional editions of Windows 10 have a "Defer Upgrades" option in the Advanced options interface. If you enable this, you'll receive security updates automatically. Windows 10 will put off downloading feature updates for several months until they've had plenty of time to be tested on home PCs. This new design makes business PCs more reliable and gives system administrators the ability to test new feature updates before they let their users get them. If you get an option to upgrade to Windows 10 Professional, you could enable this option yourself.

As to the earlier mention of the "Advanced Options" link in the Windows Update interface, you can find just two "Choose how updates are installed" options. You can pick "Automatic," which is the default and allows Windows to automatically download updates, install them, then schedule a reboot time that's convenient, or for when you aren't using your PC. Also, you can choose "Notify to schedule restart," which will prevent your device from automatically rebooting without your prior confirmation. Both way, those updates will be automatically downloaded and installed and you must control when to reboot your system.

If, and when, you ever encounter any instability after downloading updates for Windows, or any other universal Windows apps, you can easily uninstall problematic updates afterwards. Simply view your update history by opening the Windows Update interface, click "Advanced options," then click "View your update history." A list of updates will display then you can click “Uninstall updates” to get a display of the updates you can uninstall.

Another delightful change in Windows 10 Updates is that you won’t have to update your PC or device after resetting or restoring your PC or device. In previous versions, this process alone could take hours. With the new Windows 10 build, however, the PC Reset feature restores your PC with any and all updates being downloaded in the restoral process, so you won’t have to re-download all the Windows updates that have been released.

Conclusion

Windows 10 takes activation of their new build very seriously, but at least with this version you are not forced to have a specific Microsoft email address. Allowing the use of Gmail addresses and other mail servers has made it easy for all users to get their digital license registered without trying to remember an unfamiliar or obscure email address. Additionally, if you have the correct hardware on your PC or device, you can setup your log-on for your activation with a fingerprint, face recognition, and iris recognition.

It's refreshing to have the start menu back, along with its many changes, including the ability to pin the Recycle Bin to the start menu and the File Explorer. The newly designed start menu has an easy-to-navigate, traditional, and alphabetical, pop-up-style menu.

Much love to the new File Explorer which is easy to navigate. Research has found that most users report the default display mode allows for easy navigation through all local folders, including cloud drives, making it fantastic for copying and pasting, and dragging and dropping files and documents from all cloud-based applications to the local drive.

At long last, virtual desktops have made it to Windows 10! Now all of us power users can spread out apps across multiple screens. An overview of all your virtual desktops can be seen with the click of a mouse.

Cortana arrives on the desktop! Set her up and get chatty with Cortana. The better she gets to know you, the more effective she will be. She's a virtual assistant, and you will just love her.

Windows 10 nails it with a powerful look while displaying the layout of all drives in one display. At first, it seems confusing, but when you understand the layout, it's very organized and easy to navigate. By typing "storage" in the search box, you can go right there and see exactly what types of files are taking up your hard drive space.

And Windows 10 has renewed the Action Center that lives in the lower right-hand side of the desktop. It provides a stream of notifications that come in from any application installed on your system. It definitely takes the guess work out of updating your apps. Click on that little speech bubble icon in the system tray to open and configure your notifications.

Exclusive to Windows 10 is Microsoft's new, integrated browser called Microsoft Edge. You can't run it on Windows 8 or below. Only Windows 10 users can take advantage of web page annotations, reading view, and Cortana search. It's simple really; Microsoft wants you to use their services, and this user has a very high opinion of Microsoft's new browser. It updates automatically—so simple!

I remember the Windows 8 tablet mode was a disaster, but the new Windows 10 interface is much more appealing. Simply open the Action Center to switch manually to tablet mode or get out of it again.

Windows 10 integration with OneDrive is genius. It's a cloud storage for users that enables compatibility with most of your devices, including Android, iPhone, cell phones, tablets and netbooks. Now you can have access to all of your important documents that are securely stored on the cloud-based OneDrive.

Windows 10 does a great job in the latest build in presenting Settings, which displays a comprehensive and modern interface to make changes to the system. A simple search in the search bar for "Settings," or saying "Settings" to Cortana will make it display. You can also click on the start menu, then click on Settings to the lower left. This build has moved more of the key system settings that have been available to users through the Control Panel.

The new Mail and Calendar features can keep you organized and on the go. Its compatibility with Android apps and iPhone apps keep you showing up on time for important events and appointments, and allows you to have access to both from your mobile device.

Wow, hat's off to Windows with their latest enhancement of the Command Prompt. It's a bit dangerous for the average user, but it's a dream for the power user with new features and new commands.

Gamers will be excited with Windows 10's ability to stream Xb**ox One games** from one place to another. Although it isn't an original idea, the link Microsoft has built with the Xbox One and Windows 10 devices could be the best use of the Windows platform yet. If the kids want to use the big screen in your living room, you can stream your Xbox One gaming up to your laptop or desktop upstairs, too.

And finally, Windows 10 peer-to-peer updates. The best enhanced feature of all. Gone are the days of spending time or having your system lag while downloads were coming in. The new peer-to-peer updates force them on your system and download them in the background so you are always up to date. In addition, another new feature has been added to the System Reset; the updates are now automatically installed as part of the reset. So gone are the hours upon hours of downloading and installing incoming updates when you can do a system restore.

All in all, Microsoft has worked really hard listening to the customer base with its upgrades and enhancements in the latest build of Windows 10, and I'm sure it will only get better, starting with a new build version scheduled to be released in late spring of 2017.

Thank you for reading. I hope you enjoy it. I ask you to leave your honest feedback.

I think next books will also be interesting for you:

iPhone 7

C++

C++
The Ultimate Crash Course to Learning the Basics of C++
Paul Laurence

Amazon Echo

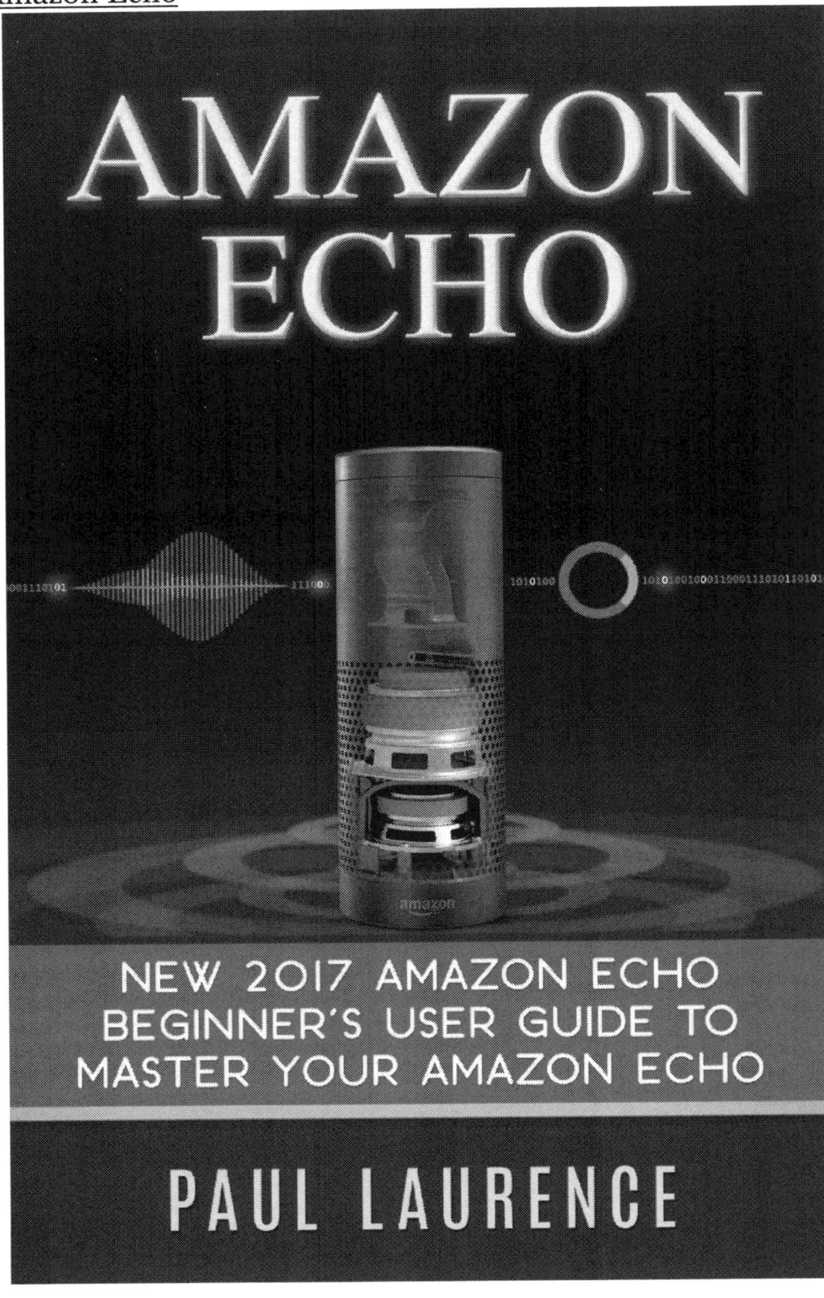
AMAZON
ECHO
NEW 2017 AMAZON ECHO
BEGINNER'S USER GUIDE TO
MASTER YOUR AMAZON ECHO
PAUL LAURENCE

44471765R10040

Made in the USA
Middletown, DE
08 June 2017